Biology of Cancer

Randall W. Phillis
University of Massachusetts

Steve Goodwin
University of Massachusetts

San Francisco Boston New York
Cape Town Hong Kong London Madrid Mexico City
Montreal Munich Paris Singapore Sydney Tokyo Toronto

Acquisitions Editor: Michele Sordi
Assistant Editor: Michael J. McArdle
Marketing Manager: Josh Frost
Manufacturing Buyer: Vivian McDougal
Designer: Detta Penna

Cover image: G. Steven Martin

ISBN 0-8053-4867-0

Copyright © 2003 Pearson Education, Inc., publishing as Benjamin Cummings, 1301 Sansome St., San Francisco, CA 94111. All rights reserved. Manufactured in the United States of America. This publication is protected by Copyright and permission should be obtained from the publisher prior to any prohibited reproduction, storage in a retrieval system, or transmission in any form or by any means, electronic, mechanical, photocopying, recording, or likewise. To obtain permission(s) to use material from this work, please submit a written request to Pearson Education, Inc., Permissions Department, 1900 E. Lake Ave., Glenview, IL 60025. For information regarding permissions, call 847/486/2635.

Many of the designations used by manufacturers and sellers to distinguish their products are claimed as trademarks. Where those designations appear in this book, and the publisher was aware of a trademark claim, the designations have been printed in initial caps or all caps.

15–CRS–10
www.aw.com/bc

Contents

Cancer 1
    Where Do Cancer Cells Come From?   3

Cancer Cells Have Profound Genetic Defects 4
    Damage Control   4
    Tumor Suppressors   7
    Growth Rates and Rapid Cell Division   8

The Cell Cycle 9
    Oncogenes—Cell Division Stimulants   12

The Six Hallmarks of Cancer Cells 14
    Self Sufficiency in Growth Signals   15
    Insensitivity to Growth Inhibitory Signals   15
    Evasion of Programmed Cell Death   16
    Limitless Replicative Potential   17
    Sustained Angiogenesis and Nutrient Supply   19
    Tissue Invasion and Metastasis   20

Evolution of Tumor Cells 20
    Side Effects and Dirty Tricks—Drug Resistance   23

Hereditary vs. Sporadic Cancer 24
    Smoking and Cancer   26

    Web Resources   29
    Books and Articles   30

Cancer

Richard is 60 years old. For the past few months he has been feeling a little under the weather. He has chronic indigestion and his stomach is somewhat sore. He is more tired than usual, and when he eats his indigestion flares up. He goes to the doctor, who prescribes an antacid but it doesn't really help. A few weeks later, he wakes up itching all over. His wife takes one look at him and is shocked to see that his skin and the whites of his eyes are yellow. His doctor is worried about his sudden jaundice and orders a CAT scan. The image of Richard's pancreas reveals a mass. The doctor gives Richard the bad news: He probably has pancreatic cancer and the prognosis is not good.

Richard undergoes surgery but the tumor has spread into tissues around the pancreas and cannot be removed. The surgeon sees the signs of malignancy from the shape, color, texture, and spreading nature of the tumor. He removes a sample of the tumor for laboratory analysis. The surgeon also corrects the blockage of his bile duct that lead to jaundice and marks the tumor with metal clips. These clips will be used in Richard's subsequent therapy to focus a beam of radiation directly at the tumor.

In the lab, the sample of Richard's tumor is stained and examined under the microscope. The pathologist also recognizes the cells as cancerous. Compared to the surrounding healthy cells, the tumor cells are more round and not well organized into layers (Figure 1, page 2). The nuclei in the tumor cells are enlarged and many of the cells are visibly undergoing mitosis. Further tests reveal that the cells are secreting unusually high levels of proteins normally produced by pancreas cells.

Richard undergoes months of chemotherapy and radiation treatments. These treatments help, and Richard feels better for several months. When additional CAT scans are performed, they show that the tumor has shrunk. But as the months pass, Richard's condition worsens. His stomach hurts more and he starts losing weight. The tumor has started to grow again despite continued chemotherapy. After a few short months, the cancer is growing out of control. Not only is the tumor in the pancreas growing, but the cancer is spreading. Richard develops new tumors in his liver, bones and lungs. His condition

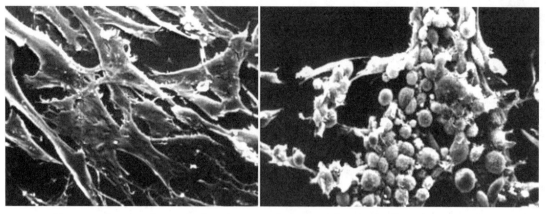

FIGURE 1. Photographs taken by a scanning electron microscope of normal cells and cancer cells. Note that the normal cells are fairly flat, have complex shapes and have many attachments to each other and the underlying growth surface. In contrast, the cancer cells are round, have lost their complex shapes and are poorly attached to each other, or anything else. *Source:* Judith Kimball's web site, http://www.ultranet.com/~jkimball/BiologyPages/C/CancerCellsInCulture.html. *Photographs:* G. Steven Martin.

rapidly deteriorates and his weight loss accelerates. When Richard finally dies, he has lost over 100 pounds.

This kind of diagnosis, treatment, and death from cancer is replayed all too often. Over 1.2 million cases of cancer will be diagnosed this year in the United States alone, and over half a million cancer deaths will occur. Pancreatic cancer will make up only 5% of those deaths, but is one of the most difficult to treat, and kills over 95% of its victims within five years of diagnosis. Cancer of the lung, breast, prostate, and colon account for the majority of the other cancer cases.

As this example shows, cancer results from the loss of control over several key cellular properties. Cells that should not divide, divide. Cells that should not be invading surrounding tissue, invade surrounding tissue. Cells that should remain in specific tissues or organs, migrate to new locations in the body and continue to grow. Cancer cells that leave their original location for a new location in the body and begin to divide in the new location are said to have metastasized. These three traits—dividing when they should not divide, having the capacity to invade surrounding tissue, and having the capacity to metastasize—separate cancer cells from normal cells.

Extensive scientific and medical resources have been mobilized to understand the basic biology of cancer and treat those afflicted by the disease. More than $50 billion will be spent this year for cancer healthcare, and nearly $15 billion for research. It is relatively easy to understand how the money is spent on

TABLE 1. American Cancer Society Data—Cancer Incidence and Deaths—2001

Males			Females		
Type	*New Cases*	*Deaths*	*Type*	*New Cases*	*Deaths*
Prostate	198,100	31,500	Breast	192,200	40,200
Lung & Bronchus	90,700	90,100	Lung & Bronchus	78,800	67,300
Colon & Rectum	67,300	27,700	Colon & Rectum	68,100	29,000
Bladder	39,200	8,300	Uterus	38,300	6,600
NH Lymphoma	31,100	13,800	NH Lymphoma	25,100	12,500
Melanoma	29,000	5,000	Ovary	23,400	13,900
Oral	20,200	5,100	Melanoma	22,400	2,800
Kidney	18,700	7,500	Bladder	15,100	4,100
Leukemia	17,700	12,000	Pancreas	15,000	14,800
Pancreas	14,200	14,100	Thyroid	14,900	800

TABLE 1. Cancer incidence by type. *Source:* "Cancer Facts and Figures 2001." American Cancer Society. http://www.cancer.org/eprise/main/docroot/STT/stt_0_2001?sitearea+STT&level+1.

cancer treatment. Healthcare is expensive, and the array of medicines, physicians, nurses, facilities, and sophisticated medical equipment account for this expense. But what about the research money? What do scientists do when they are studying cancer? What have they discovered about how the biology of cancer cells is different from normal cells? How will those discoveries be translated into treatments to help patients like Richard?

WHERE DO CANCER CELLS COME FROM?

In some ways, cancer behaves like an infection. It starts at a specific place, and then spreads to other places in the body. In bacterial infections, bacterial cells come in contact with the body and start to grow. Bacteria grow by dividing, and as they divide a cluster of cells is formed, called a colony. Colonies of bacterial cells can break up and the cells that are released can spread and start to grow at new sites and form new colonies. Some cells from a bacterial colony can travel in the blood or other body fluids and grow at sites distant from the initial infection.

There is a critical difference between bacterial infections and cancer. Cancer is produced not from cells that originate outside the body, but from changes in cells that are part of the body itself. Richard had pancreatic cancer. Other patients may have lung cancer or colon cancer or breast cancer. The assignment of these different types of cancer reflects the cells types or organs in which the cancer started. When cancer spreads, it is these abnormal body cells, produced from rapid cell division, that break free and spread through the body fluids to distant sites where they lodge in healthy tissues and continue to divide. Because these cells are actually the body's own cells, they are not recognized as foreign, infectious agents and are not aggressively attacked and removed by the immune system.

If cancer cells are not foreign cells, but are derived from normal cells, what happened to them to make them change? Why are they abnormal? What do they do "wrong"?

Cancer Cells Have Profound Genetic Defects

To start this discussion, it is important to look more closely at cancer cells and their defects. When Richard's tumor cells were sent to the lab, they were subjected to several tests, one of which was to determine their karyotype. This is a microscopic analysis of the cells to determine how many chromosomes they have and whether the chromosomes are normal or have defects that can be seen under the microscope. Richard's tumor cells had striking changes in their chromosomes (Figure 2). Many of the chromosome were broken or had parts missing. Others had extra parts attached that came from other chromosomes. Several extra copies of some chromosomes were present. Other chromosomes were nearly missing, with only a part of a single copy remaining. These are startling defects. No normal cell ever has such problems. How do cancer cells develop these problems? How are these genetic changes related to the abnormal behavior of cancer cells? How do cancer cells survive despite having such problems?

DAMAGE CONTROL

To understand how cancer cells accumulate such high levels of genetic and cellular damage, we need to understand what normal cells do in the face of significant damage to their chromosomes. Normal cells have a robust system for damage control that can do four things: (1) Detect cellular damage, especially DNA damage; (2) arrest cell division and prevent the replication of damaged cells; (3) activate damage repair systems; and (4) activate cell death if damage cannot be repaired. This system of damage control functions routinely in normal cells. For example, when you get sunburned, your skin cells activate damage control, stop dividing and activate damage repair. Skin cells mildly damaged by solar radiation can be repaired and when repair is complete, the cells start dividing again. However, some cells may be damaged so severely that the damage cannot be repaired, and a system in the damaged cell is activated that caus-

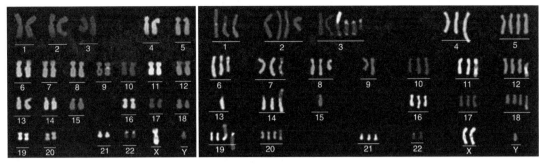

(a) Normal (b) Cancer karyotype

FIGURE 2. Karyotypes of a normal cell and a cancer cell. Note that there are two copies of each of the 22 chromosomes and an X and Y. The staining technique used to obtain these karyotypes allows each different chromosome to be reocognized because it stains a different color. Note that in the cancer karyotype, there are many extra chromosomes present. Some of the chromosomes also have extra DNA attached such as the right most copy of chromosome 4. *Source:* http://www.nature.com/genomics/human/slide-show/4.html.

es cellular suicide. This process, called apoptosis or programmed cell death, is a critical part of damage control. Cells damaged beyond repair normally remove themselves by activating cell death systems. You might think that cells so badly damaged would just die anyway. The problem is that cells that die directly from damage or injury, called necrotic cell death, are not effectively removed from surrounding tissue. Necrotic cell death triggers inflammation and the leftover cell debris can be a target for bacterial infection. In apoptotic cell death, the cell is efficiently disassembled and the parts are readily engulfed by healthy cells and recycled.

To think through the logic of damage control in another way, let's think of cells as automobiles. When our car is damaged, we quit driving it and take it to the repair shop. The mechanics then fix the car or decide the damage is too extensive to fix, and have the "totaled" car hauled off to the junkyard. Most damage to cars hinders their performance. A fender may be dented and rub against a wheel or the radiator may leak and cause the engine to overheat and not run well. Most damage to cells is similar. It prevents normal cellular function and needs to be repaired for cells to start working properly again.

What is the damage in cancer cells like? Cancer cells divide out of control and can spread throughout the body. Our automobile model of damage and repair makes sense if we consider the idea that damage to cancer cells is more complex than a dented fender is to a car. Let's propose that cancer is to cells what driving out of control is to cars. What sorts of damage could cause cars to

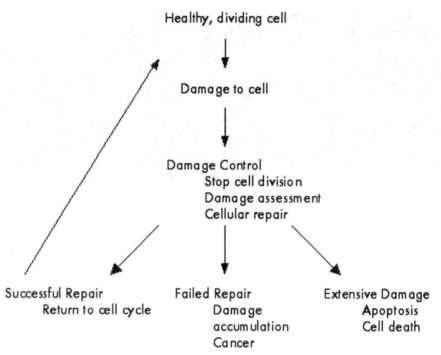

FIGURE 3. Normal cells divide under proper control but can be damaged from toxins or radiation. When damage occurs, the damage control systems are activated. First, damage is detected by several proteins that become active when damage to DNA or other cellular components occurs. These damage detectors then activate damage control genes such as p53 that stop the cell cycle and activate repair. If the damage is mild, it is repaired properly and the cell becomes a normal dividing cell again. If the damage is very severe it triggers the programmed cell death system and causes cell death, called apoptosis. Finally, the damage repair systems may fail, and damage may accumulate without either being repaired properly or triggering cell death. In this case the accumulation of damage to genes can lead to cancer.

go out of control and careen around corners and crash into things? Well, what if there is damage to the brakes? What if the steering doesn't work or the accelerator gets stuck? These kinds of damage affect the car's control systems and are especially dangerous. Damage to these systems must be recognized by the driver and fixed by a mechanic or the car will go out of control when driven. If the problem is too extensive to be repaired, the car should be taken to the junkyard.

Let us take this analogy and apply it back to damage control systems in cells (Figure 3). To activate damage controls systems, the damage must be detected and a decision must be made by the cell to stop dividing and initiate repair. Several genes that encode proteins that function in these systems have been identified, but the details of many parts of this system are still unresolved. Damage to DNA is detected by proteins that bind to chemically altered nucleotides or broken DNA molecules. Damage to membranes and proteins can also be detected by several enzymes normally present in cells. These damage detectors function much like a driver sensing something is wrong with a car, feeling the brakes not work correctly or the engine running improperly.

When damage is detected, a coordinator of damage repair is activated. This job is similar to the manager of a repair shop. This individual has to assess the extent of the damage and make sure the repairs are completed properly. If the damage is too extensive, this manager must call the junkyard and have the car hauled away. In cells, the job of damage control manager is performed in part by a key protein in the cell called p53. The p53 protein has many functions including arresting cell division, activating repair enzymes, and, if necessary, triggering cell death.

The gene encoding p53 is mutant in over half of all cancers. When this mutation occurs, damage responses are much more likely to fail. When p53 does not function correctly, repair systems are not properly activated, and most importantly, cell death is not triggered when cells have severe damage and should be destroyed. If the management of damage control is defective, cellular defects go unrepaired. This failed repair is especially dangerous if the defects in the cell lead to increases in cell division rates, increased ability of cells to penetrate tissues by invasive growth, or the ability of cells to break free and spread to distant sites in the body. When these defects are present and not repaired, the result is an aggressive cancer that is rapidly growing and likely to spread.

TUMOR SUPPRESSORS

The genes that function in the damage control systems fix cells when they are broken. In cancer cells, these genes are mutant, and cannot properly perform their tasks. The defects in these genes reduce their expression or block their ability to do their jobs, and are called loss of function mutations. Genes that contribute to cancer formation when they experience loss of function mutations are called tumor suppressor genes. They encode proteins whose normal job is to suppress the changes in cells that lead to cancerous growth. Some tumor suppressor genes encode proteins that detect DNA damage; others encode proteins that function in the repair of damage to DNA or other cellular components such as proteins or lipids. Still other tumor suppressor genes, such as TP53 that encodes the p53 protein, function as coordinators of repair systems. They stop cell division when damage is detected and activate repair systems. Finally, a group of tumor suppressor genes function in programmed cell death by helping cells commit suicide when the cells have severe damage that cannot be properly repaired.

It is not hard to see how loss of tumor suppressor functions could contribute to cancer. Cells with mutations in tumor suppressor genes accumulate damage that is not repaired. If the damage affects key control systems, it could lead to cancer formation. As mutations build up in cells, they would lose their mature shapes and physiology. This process, called dedifferentiation, is characteristic of cancer cells, and was observed in Richard's tumor cells. His cancer cells didn't look normal to the surgeon, nor the pathologist in the lab. They had lost their shape and the tissues they formed had lost normal organization and structure. In normal circumstances, these cells die from apoptosis and are lost from the body, but cancer cells are frequently incapable of apoptosis and persist and spread despite their damage and malfunctions.

GROWTH RATES AND RAPID CELL DIVISION

We have discussed how cancer cells are defective and have defective damage control systems, but what about their growth? Why do tumors made of cancer cells get bigger? What changes occur that cause increased growth rate? Normal tissues and organs have controls that regulate their growth. When they achieve proper size, their growth stops. These controls on growth are ignored or lost in tumor cells.

To start our discussion of cancer cell growth, we have to distinguish between two basic possibilities of tissue growth. One possibility is that cells increase in size and content, and bigger cells make bigger tissues. The other possibility is that growth results from an increase in cell numbers. In cancer more cells are formed, but the cells are not appreciably larger than normal cells.

Microscopic analysis of tumors allows us to distinguish between these possibilities. When Richard's tumor was examined, the cancer cells were visibly different from surrounding healthy cells, and there were clearly many of them. In fact, the cancer cells themselves tended to be somewhat smaller than normal cells. The cancer cells had dedifferentiated, and lost their mature characteristics and their complex shapes had been lost. Richard's cancer resulted in the growth of tumors and increased amount of tumor tissue, but the cells that made up the tumors were smaller than normal cells. This occurred because tumor growth was due to excess cell division and increased cell numbers rather than an increase in cell size. Even though the cancer cells within the tumor were smaller than normal cells, the very large number of the cancer cells accounted for the increase in tumor size.

To understand the reasons for increased rates of cell division in cancer we must think about how cell division is regulated in normal cells. We have to look for special kinds of defects that could make cells divide too fast. To bring up our car analogy, most damage to a car will make it run very poorly. However, having a stuck accelerator is a specific defect in a key control system that would make the car go too fast. Many of the defects in cancer cells are in key control systems that accelerate cell division.

The Cell Cycle

Many conditions must be met before a cell can divide normally. Some of these are internal conditions having to do with cell growth and metabolism. A cell must grow in size and content before it divides (Figure 4 on page 10). It must produce enough protein, membrane, organelles, and carbohydrates so that when those materials are divided in half in cell division, there is enough to support each of the two daughter cells formed. A cell must also fully replicate its DNA, so a complete genome can be provided to each daughter cell. The status of these aspects of internal growth are monitored by regulatory systems that block cell division unless proper growth has occurred.

Other conditions are external and involve signals from other cells. External conditions required for proper cell division are more easily understood if a cell is viewed in a "social" context, where it is part of a community of cells that must cooperate to form an organism. For cells to contribute to tissues or organs, they must be in contact with the appropriate neighboring cells and be properly anchored in place within the tissue. Cells have a complex array of signaling systems that relay information about contact with other cells and the strength of the anchors that hold the cell in place. Cells must also receive signals from distant parts of the body that coordinate the function of different organs. These signals are transmitted by hormones and growth factors released by other cells and endocrine glands. These signals allow coordination of growth among different tissues. Regulatory systems sense these external signals and coordinate the information to regulate cell division.

It is important to understand that these growth signals can be either positive or negative. Many signals stimulate division and are required for growth of tissues. However, many of the signals cells inhibit or block cell division and are an important part of growth control. Most cells in the body are not dividing much at all and many cell types have undergone terminal differentiation and matured, never to divide again. The signals these mature cells receive from neighboring cells, hormonal control systems, and internal triggers prevent them from returning to a cycling state of division.

The signals controlling growth, both positive and negative, are integrated in a basic control system regulating the cell cycle. The cell cycle refers to the sequence of events that are repeated each time a cell grows and divides to form two daughter cells. As the cycle proceeds, the cells goes through a growth phase

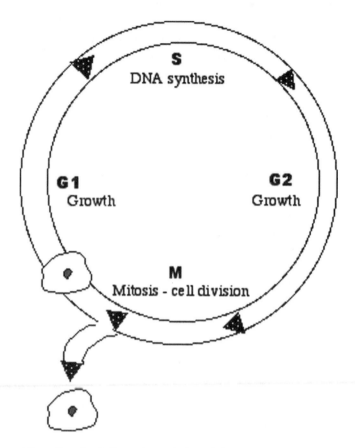

FIGURE 4. The cell cycle. Cells go through four different stages each time they divide. Cells start with a growth phase, called G1, in which they synthesize proteins, lipids, and carbohydrates and perform important metabolic functions. Once they grow properly and receive signals that stimulate cell division, they move to S phase, in which they replicate their DNA. After DNA replication, further growth and metabolism occurs. Finally, the cells undergoes mitosis, or M phase, in which it divides its chromosomes evenly and splits into two genetically identical cells. Each of those cells can then cycle, or go through the process again.

called G1, a DNA synthesis phase called S, another growth phase, called G2, and finally into mitosis, or M phase when the cell divides.

The transitions from each phase of the cell cycle to the next are tightly regulated by a system that combines signals about both internal and external cellular cues into a single control switch for cell division. This system involves a series of proteins called cyclins that act as triggers for progression through the cell cycle (Figure 5). At the start of the cell cycle, the concentration of cyclins is

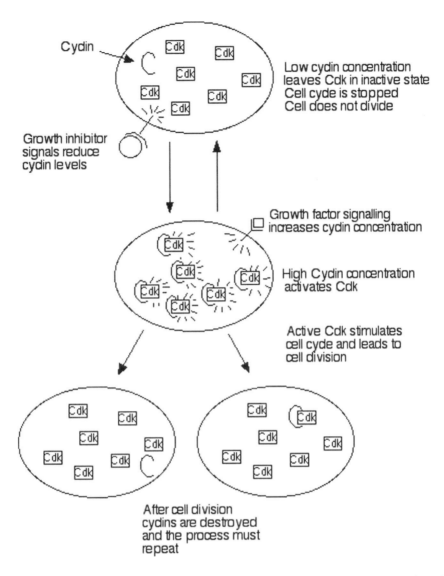

FIGURE 5. Early in the cell cycle, cyclins are at very low concentrations in the cell. When growth factors signal the cell, it stimulates cyclin production. When the cyclins are present in high concentration, they bind with the cdk's, or cyclin-dependent kinases. These two-molecule combinations, or dimers, then signal other molecules that allow the cell cycle to proceed. Once the cell cycle proceeds, the cyclins that were at high concentrations are uncoupled from the cyclin-dependent kinases are destroyed, and the cyclin-dependent kinases are deactivated.

very low. As signals are received to stimulate cell division, cyclins accumulate to high concentrations within the cell. When cyclin concentration increases, they combine with a second class of proteins called cyclin-dependent kinases. The dimers of cyclins and their corresponding cyclin-dependent kinases become active enzymes that stimulate additional proteins. These target proteins function to actually move the cell from one phase of the cell cycle to the next. For example, many of the targets of cyclin-dependent kinases activated at the end of the G1 phase are enzymes required for DNA synthesis. Thus, the DNA replication machinery used in S phase is engaged actively only after all of the signals from G1 phase cues have combined to produce high G1 cyclin levels. These integrated signals are combined into a unified activation event triggered by the G1 cyclin-cyclin-dependent kinase enzyme complexes.

Each of the transition steps of the cell cycle, from G1 to S phase, from G2 to M phase, and the final completion of mitosis are regulated and triggered by specific sets of cyclins and their corresponding kinases. The production of cyclins at each stage are stimulated by positive growth signals and inhibited by negative signals. Through this give-and-take, the cell cycle can be either stimulated to allow the cell to divide, or stalled, arresting the cell in its existing growth phase.

ONCOGENES—CELL DIVISION STIMULANTS

Early in the study of genetic changes in cancer cells, it was found that some viral infections could transform normal cells into cancer cells. This transformation was traced back to specific genes carried into the cells on the viral DNA. The *src* oncogene (pronounced "sark") represents one of the first of these transforming genes discovered by Harold Varmus and Michael Bishop. These researchers were eventually awarded the Nobel Prize for the discovery of this gene in a virus, and the identification of a corresponding gene normally present in the genome of healthy cells. They proposed that oncogenes, genes that can transform normal cells into cancer cells, were just mutant forms of genes normally present in cells. The normal counterparts, called proto-oncogenes, function in regulating the cell cycle. When those genes become overactive, they accelerate the cell cycle by stimulating cyclin production and produce rapidly dividing cells. The functions of many proto-oncogenes are now reasonably well understood. Many proto-oncogenes function in the signaling systems from internal or external growth cues. For example, the *src* proto-oncogene functions to provide information about how well cells are anchored in place, and provides positive growth signals when anchors are intact.

Most oncogenes are caused by gain of function mutations in proto-oncogenes (Figure 6). This kind of mutation does what it sounds like it does. It causes genes to gain functions, such as being expressed too much, or in cells where

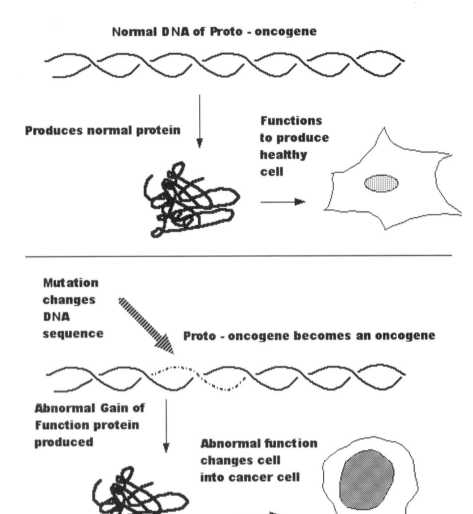

FIGURE 6. Proto-oncogenes are normal genes that perform regular functions within the cell. When proto-oncogenes function, they contribute to the normal growth and regulation of cellular functions. Mutations can occur in the DNA sequence of proto-oncogenes to turn them into oncogenes. When this occurs, the protein they produce does not function normally, and gains new functions. In turn, it leads to changes in cell growth and metabolism that contribute to cancer.

it is normally turned off. Many of these mutations affect proteins that function to transmit signals from growth factors and cause them to be turned on constantly. In this case, the cell thinks that there are growth factors present even when there are not. The cell responds by building up its cyclins and speeding up the cell cycle, even though the signals to do so are not actually present.

Now let us imagine the worst-case scenario and consider a cell that has oncogene mutations and tumor suppressor mutations. Such a cell would have abnormal function in both genes that control the cell cycle and the rate of cell division, and genes that regulate damage control. What would result? First, the loss of damage control would lead to the accumulation of cellular damage, especially genetic mutations. When these mutations occur in proto-oncogenes, they generate oncogenes that speed up the cell cycle. Further mutations, again still not repaired by damage control, might increase cell motility. Now the rapidly dividing cells would be able to move more rapidly and spread from their original site. As more mutations accumulate, more oncogenes are created and the cells would divide more and more rapidly. In addition, further mutations can occur in the damage control system and make it even worse. It is this disastrous downward spiral caused by the combination of tumor suppressor mutations that block proper damage control, and oncogene accumulation that speeds up cell division and increases the cell's invasive properties that is characteristic of advanced cancer.

The Six Hallmarks of Cancer Cells

Douglas Hanahan and Robert Weinberg recently wrote a review of cancer research in which they outlined six features common to all types of cancer cells. This review and the ideas it contains are an excellent synthesis of many complex and diverse discoveries in cancer research. The six features of cancer cells Hanahan and Weinberg (2000) consider are:

1. self-sufficiency in growth signals or response

2. insensitivity to growth inhibitory (antigrowth) signals

3. evasion of programmed cell death (apoptosis)

4. limitless replicative potential

5. sustained angiogenesis [stimulation of blood vessel growth]

6. tissue invasion and metastasis.

These specific features of cancer cells are the result of changes in cell physiology produced by mutation and genomic alteration characteristic of cancer cells. The specific genes involved and the manner in which they function are only partially understood. Understanding what these genes do and how their malfunction contributes to cancer is the subject of a wide array of intense research efforts. Given the general outlines of the cancer process already discussed, we will now explore some of the important concepts underlying each of these "hallmarks of cancer."

SELF SUFFICIENCY IN GROWTH SIGNALS

Let us start with external signals controlling growth. Several tumors overactivate receptor systems that sense growth factors and regulate cell division. Some cancer cells achieve this by increasing the number of growth factor receptors on their surface so that even small amounts of growth factor produce a strong signal response inside the cell and activate cell division. Other cancer cells have mutant receptors that transmit their signals even without growth factor present. In a similar way some cancers have mutations in the proteins inside the cell that are activated by the receptor and transduce or transmit the signal into the cell. These mutant oncogenes activate cell division and are common in many cancer cells. An example is a gene called *ras*, which encodes a protein that functions as a signal transduction molecule. The mutant *ras* protein becomes overactive in cancer cells and delivers a growth signal even when none is actually being transmitted from outside the cell.

Among the most devious of the systems that lead to the mistaken signaling of growth factor systems are cancers that co-opt neighboring healthy cells to overexpress growth factors. This effectively represents a hijacking of healthy cells by cancer cells to support their rapid growth. Cancer cells that thrive on extra growth signals get their "fix" by releasing molecules that affect nearby cells and fool them into expressing high levels of growth factors. With this in mind, the complexity of cell types within tumors can make sense. Many tumors are a mixture of cancer cells that are capable of rapid growth, invasive growth and metastasis, and noncancer cells that may function to "support" the rapid growth of the cancer cells.

INSENSITIVITY TO GROWTH INHIBITORY SIGNALS

Just as cancer cells can become overly sensitive to growth factors, cancer cells can also become selectively deaf to signals that normally slow cell division. Most of these changes involve mutations that block receptors for growth inhibitory signals. An interesting example is found in the growth inhibiting signals transmitted when cells are properly anchored in place. The anchors that attach cells to each other are formed by special cell adhesion molecules

(CAMs), which are imbedded in the membranes of adjacent cells and stick to each other. When such contact is made, the internal parts of these molecules activate a signaling system that reduces cyclin levels and slows or stops cell division. A key pair of players in this system are molecules called cadherins, which are the transmembrane anchoring molecules, and catenins, which are the internal signaling partners of cadherins. Defects in these systems can lead to the loss of the stop signals associated with cell anchors and cause unregulated cell growth.

EVASION OF PROGRAMMED CELL DEATH

Many of the strategies cancer cells use to sustain rapid growth and spread throughout the body cause cellular damage and would be suicidal for normal cells. First of all, most changes in cancer cells result from some kind of mutation. Mutational damage to DNA is detected by normal cells and frequently triggers cell death. Cancer cells accumulate an impressive array of mutational defects but fail to trigger cell death when the mutations occur. In addition, the metabolic stresses of rapid cell division, including oxygen deprivation, normally trigger cell death, except in cancer cells. Finally, cell death is triggered when cells lose contact with other cells and their anchors are disrupted, except in cancer cells. The biochemical pathways that trigger apoptosis or cell death are becoming more clearly understood and are quite complex. There are several mutations, including in the p53 encoding gene as described earlier, that can disrupt this process and allow cancer cells to survive despite the presence of several apoptosis-triggering signals.

The loss of apoptotic capacity in cancer cells has important implications for cancer treatment. Many cancer treatments in wide use today, including radiation and chemotherapy, are designed to cause DNA damage to rapidly dividing cells. The logic of these treatments is that targeting DNA synthesis will focus the treatment on cells undergoing rapid division. Therefore, cancer cells will be selectively affected, and slower growing healthy cells will experience less damage. The catch is that even though the rapidly dividing cancer cells accumulate damage from the treatment, they frequently fail to react to that damage by triggering cell death. Therefore, the very design of the treatment is often defeated by one of the properties that comprise the hallmarks of cancer.

The good news is that research reveals the mechanisms of apoptosis are quite complex and can be triggered many different ways. Though some cancers may have mutations that block apoptosis from damage to DNA, they may still activate cell death from loss of cellular anchors. A number of cancer treatments are in development target the apoptosis pathways that are still intact in cancer cells, to activate cell death to eliminate cancer cells.

LIMITLESS REPLICATIVE POTENTIAL

Almost all normal cells can divide only a limited number of times. For many years it has been known that cells removed from the body and grown in cell culture survive for a limited number of cell divisions. Once human cells have divided 60–70 times, they become incapable of further division and die. In contrast, cancer cells never stop dividing. They can be cultured in the laboratory indefinitely. This raises the question, what exactly wears out in normal cells that stops their ability to divide?

Much of the answer is found in the special structures at the ends of chromosomes, called telomeres. These are special repeating DNA sequences that serve two critical functions. First, they help protect the free ends of chromosomes. Double-stranded ends of DNA molecules are viewed by cells with deep suspicion. Except for telomeres, the only source of double-stranded DNA ends in cells would be from broken chromosomes, or DNA introduced by viruses—both clear signs of trouble. When double-stranded DNA ends do occur in cells, several proteins bind to them and check them for matching sequences elsewhere in the genome. If such matches are found, the ends are joined in DNA repair. If there is no match, enzymes are employed to degrade the DNA or even kill the cell. Telomere sequences are protected by special telomere specific proteins that bind to their repetitive DNA sequences. This combination of repetitive sequence and protective proteins allow the normal ends of chromosomes to exist in cells without triggering repair.

A second important function of telomere sequences is to accommodate problems with DNA replication at double-stranded DNA ends. DNA polymerase, the enzyme that functions in DNA synthesis, can add new DNA subunits only to the ends of an existing DNA strand. DNA molecules are double stranded, and the strands run in opposite directions. During DNA replication, one strand can be synthesized to the end of a chromosome because the DNA strand is oriented so that DNA polymerase can move out toward the end as it functions. In contrast, the other strand of the DNA molecule is oriented the opposite way, and DNA polymerase moves in from the end (Figure 7 on page 18). The DNA bases right at the end cannot be properly replicated on that strand. Consequently, at each cell division the ends of DNA molecules are not completely replicated, and some sequence is lost. During each cell division, some of the repeated telomere DNA sequences are lost from chromosomes ends. The repetitive nature of telomeres accommodates this under-replication problem. At birth, our telomeres are quite long, and there are plenty of repeats to lose. However, as we age and cells continue to divide, the sequences are progressively lost until all of the repeats are gone. These "worn out" telomeres then lose their ability to bind to the protective proteins and are attacked by the repair

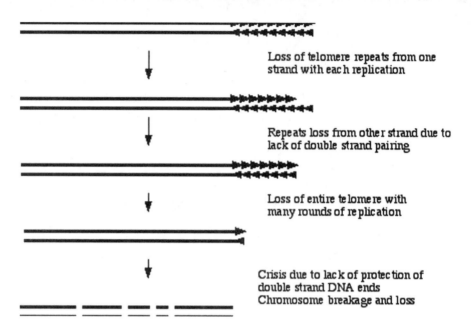

FIGURE 7. Normal chromosome ends, called telomeres, have repeated sequences. At each round of replication, a few of these repeated sequences are not properly replicated and are lost. After many cell divisions, each with a round of DNA replication, all the telomere repeats can be lost. When this occurs, the DNA ends are not protected, the chromosome enters a crisis of joining to other chromosomes, fails to separate properly in mitosis, and ultimately dies.

functions that frequently attempt join them together. This creates a long string of connected chromosomes that cannot separate properly in mitosis. When these form, the cells enters a crisis phase and ultimately dies.

Cancer cells avoid this problem. They activate a special enzyme, called telomerase, that can rebuild telomere sequences and prevent the cell from ever entering crisis. The expression of telomerase is normally limited to very few cells in the body, primarily stem cells and germ line cells in the gonads. These special cells are the rare examples of healthy cells in the body that have limitless replicative potential. Cancer cells use the strategy of these cell types and activate telomerase to continuously rebuild their telomeres and avoid the crisis and death caused by telomere loss. Indeed, nearly 90% of cancers have activated telomerase as one of the genetic changes that have led to their cancerous state.

The telomerase activity present in cancer cells has recently become the target of new experimental cancer therapies. The strategy is to block the activity of telomerase and prevent the rebuilding of telomeres. Because of their rapid rates of DNA replication, cancer cells without telomerase would quickly exhaust their supply of telomere sequences and enter crisis phase.

These therapies have some promise, but also some concerns. Most healthy cells do not have active telomerase, while most cancer cells do. This provides a mechanism for specificity of treatment. If telomerase is inhibited by some drug or genetic therapy, most healthy cells wouldn't be affected, while cancer cells would. However, the healthy cells that do express telomerase are very important. For example, the stem cells in bone marrow that divide continuously to produce blood cells have active telomerase. The potential to negatively affect stem cells with telomerase therapy has led to careful consideration of this approach and efforts to target the telomerase treatment away from such sensitive cells.

SUSTAINED ANGIOGENESIS AND NUTRIENT SUPPLY

As tumors grow and their cell division rate increases, their metabolic needs increase as well. Not only are nutrients required that build cellular structures, but oxygen is required for cellular respiration. To meet these demands, cancer cells develop the ability to stimulate the growth of blood vessels, called angiogenesis. Cancer cells release factors that activate signaling systems that cause blood vessels to branch and grow. An example is the overproduction by some cancer cells of vascular epithelial growth factor (VEGF), a secreted molecule that is released by cancer cells and interacts with receptors on the epithelial cells of blood vessels. This interaction causes the cells to grow and divide. In addition, these activated blood vessel cells become motile, and can grow into new tissues. As they do, they form new blood vessels that supply nutrients and oxygen. These functions are characteristic of normal blood vessel growth, where the cells of small vessels or capillaries divide and grow into new tissue, ultimately organizing into the tubes that are connected to existing blood vessels and extending the supply of blood. The difference with tumors is that the signals that stimulate this growth are expressed at very high levels and lead to the hypervascularization of tumors and rich supplies of blood and nutrients. Not only does this supply the cancer cells with nutrients and oxygen they require for their rapid growth, but in advanced cancers, these tumors can rob surrounding healthy tissue of adequate nutrients and oxygen and contribute significantly to the pathology of cancer.

Researchers are developing therapies to block angiogenesis stimulated by tumors and to rob them of their blood supply. For example, a therapy that would interfere with VEGF signaling by pancreatic tumor cells has been successful in mouse model systems and is currently being developed as a human therapy. Pancreatic cancers are particularly angiogenic, and it accounts in part for their aggressive growth, spread, and ability to affect overall metabolism.

TISSUE INVASION AND METASTASIS

Careful analysis shows that the spread of cancer not only involves the release from anchors or attachments cells normally have, but also requires the activation of cell motility systems. Many cells can crawl or move from place to place by rearranging their cytoskeleton. Motility is characteristic of many cells early in development. Mesenchymal cells are a key example. These cells have very active cytoskeletal systems and can send out projections, called filopodia, to explore their surroundings and ruffle their edges, forming broad protrusions called lamellipodia. Active filopodia and lamellipodia are characteristic of motile cells that can crawl or move from place to place. Some normal cells, like white blood cells, have this mobile nature as well, but most mature cells do not, and stay anchored in place. Key genes involved in mobilizing filopodia and lamellipodia are *rac*, *rho*, and *cdc42*. Misregulation of these genes is common in cancer cells capable of invasive growth and metastasis (Figure 8). Cancer cells shift from their role as stationary cells within organs to motile cells that can crawl to new locations and penetrate other tissues by changing the dynamics of their internal structural supports.

Evolution of Tumor Cells

Many genetic changes occur during the transition normal cells make to become cancer cells. The receptor systems involved in growth signaling are disrupted, cell death pathways are blocked, cell motility proteins are altered, the ability to induce changes in other cells, especially blood vessels, is acquired, and immortality is achieved by protecting chromosome ends. Each of these changes requires mutations to alter the genes that function in these central aspects of cell physiology. By the time cancer has progressed to a lethal level, an entire array of mutations has occurred to produce the hallmarks of cancer that appear in the ultimate tumor cells. But how does this progression occur? Is there an inevitable deterioration of cancer cells that lead them to this life-threatening state?

It is helpful to compare the development of cancer to the process of evolution by natural selection. In natural selection, genetic mutations that increase reproductive success become more frequent in a population over time. When this logic is applied to populations of cancer cells, it can explain the progression of small, slow-growing, localized tumors to large, rapidly growing tumors with rich blood supplies and the capacity to spread throughout the body.

A critical requirement for the selection process is a source of genetic variation. In the body, cells all start out with the same genetic makeup, and therefore no genetic variation exists. Cells become genetically different when mutations

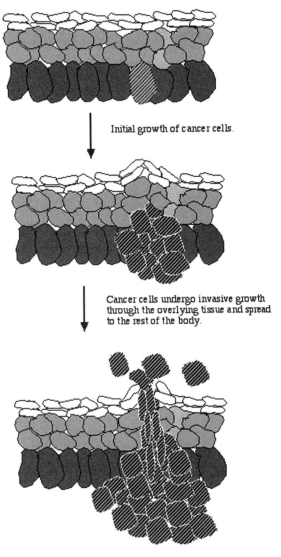

FIGURE 8. Invasive growth of cancer cells. Early in tumor development, cancer cells grow, but remain in place in the tissue. As cancer progresses, the cells develop the ability to change shape, secrete enzymes that break down surrounding tissues, and grow through adjacent cell layers. This invasive growth allows cancer cells to metastasize, or spread, to new locations in the body where they can attach and form new tumors.

occur in some cells but not others. When mutations occur at a high enough rate, it can create genetic variation among cells to allow selection to operate. Through the process of selection, cells that can divide more rapidly than others become more frequent than their slow-growing counterparts. Importantly, because mutations that occur in a cell are passed to both daughter cells at each cell division, mutations that cause rapid cell division accumulate in rapidly growing cells. As mutations occur that cause some cells to divide and grow more rapidly than others, those mutant cells would become more prevalent in a tissue. Further mutation may occur, and again this causes some of those more rapidly growing cells to divide even faster. As this process recurs, the most common cells in the tissue would be the cells that have accumulated mutations that allow them to divide the fastest. Ultimately, mutations accumulate and selection occurs until cancer cells have acquired the six hallmarks of cancer previously described. It is clear that agents that cause mutation, such as radiation and some toxins, increase cancer rates, and that the likelihood of developing cancer from these sources is correlated with their mutagenic strength.

If we follow a cell lineage from its healthy start to its final fate as a part of a fully developed tumor we might observe the following mutation sequence. Initially, a cell might acquire a mutation in growth factor receptors that would increase the balance of growth stimulating signals and decrease growth inhibiting signals. The descendants of this cell would now divide faster than neighbors and form a small benign tumor. Next, a cell within this tumor might develop an additional mutation that blocks apoptosis so that the stress of increased growth rates would not trigger cell death. Cells derived from this mutant cell would now not only be dividing faster, but would not undergo apoptosis and disappear from the tumor. Within this cell-death-resistant population of tumor cells, a mutation might occur that stimulates the release of VEGF and cause the formation of blood vessels that would supply the tumor with nutrients. Again, further mutations could occur in a cell within the tumor that activate cell motility and allow that cell and its descendants to lose anchors and move from the original site. When these cells move to new sites in the body, they would retain all of these mutations: They would be capable of growing rapidly, resistant to cell death, and able to induce the formation of new blood vessels. Finally, cells in this population could mutate further to activate telomerase and become immortal, unconstrained by the limits on DNA replication imposed by telomere shortening.

It should be noted that each mutation did not occur in every cell within the tumor. The cell in which it did occur, however, had a growth advantage, and because all of its descendants also carried the mutation, they too shared this advantage. Over time, this higher rate of growth in this population of mutant cells allowed them to outnumber other cells within the tumor that did not have the mutation. Also, a series of mutations is required. There are two logical out-

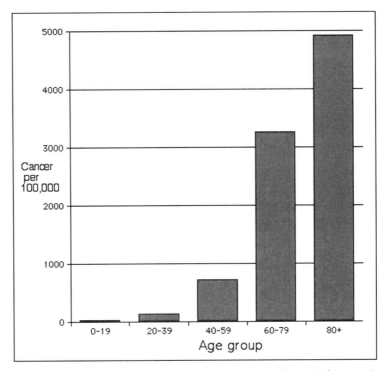

Figure 9. The incidence of cancer in different age groups in the United States. Cancer is quite rare prior to age 40, before a sufficient number of mutations have accumulated in cells of the body. However, as we age, the mutations accumulate and the incidence of cancer increases dramatically. *Source:* "Cancer Facts and Figures 2001." American Cancer Society. http://www.cancer.org/eprise/main/docroot/STT/stt_0_2001?sitearea +STT&level+1.

comes of this. First, if mutations are rare, then the accumulation of multiple mutations must take time. This is consistent with the observation that cancer is largely a disease of old age. Data on almost all types of cancer suggest that time is required for mutations to accumulate and full-blown cancer to develop (Figure 9). But age is not the only factor. Agents such as radiation or chemical mutagens that increase the rate at which mutations occur will also accelerate the appearance of cancer in the body.

SIDE EFFECTS AND DIRTY TRICKS—DRUG RESISTANCE

As cancer cells accumulate mutations, some of the changes they cause contribute to the severity of the disease and the difficulty in its treatment. For example, Richard's cancer ultimately became resistant to chemotherapy, and spread in the face of aggressive treatment. This kind of drug resistance has been shown to result from mutations that occur in tumor cells. There are several kinds of mutation that can confer resistance to drugs, but two examples are especially instructive. The first is P-glycoprotein, a transmembrane protein that can function as a pump to eject cancer drugs from the cell before the drug can have therapeutic effects. As expression of the P-glycoprotein gene increases in cancer cells, they increase their capacity to pump chemotherapy drugs out, and this dramatically decreases treatment effectiveness.

The second drug resistance feature shared by many cancer cell centers around the anti-apoptotic characteristic of cancer. Many chemotherapy agents use the strategy of damaging DNA, and rely on the DNA damage detection system in cells to activate apoptosis. However, many cancer cells experience mutations in these very systems during their development. They may no longer be able to detect DNA damage or use signals from DNA damage detection to activate apoptosis. This represents a central paradox of cancer therapy, that the very characteristics that are defined as hallmarks of cancer cells also defeat the treatment strategies commonly used in cancer therapy.

Cachexia. One of the life-threatening side effects of many kinds of cancer is uncontrollable weight loss, specifically marked by the wasting away of muscle and other lean tissue. This condition, termed cachexia, is the direct cause of 10–20% of cancer deaths. The causes of cachexia are complex and not fully understood. It is clear, however, that the metabolism of healthy cells is affected by signals released by cancer cells. Some of these signals include protein and lipids that affect appetite control centers in the brain. Other signals can directly induce metabolic changes within healthy cells. These metabolic changes then lead to loss of healthy cells, and the wasting of healthy tissue. Tumor-derived signals released in the bloodstream are received by muscle and change the rate of protein degradation within muscle cells. This capacity of tumor cells to not only elude intrinsic damage-monitoring systems, avoid cell death, hijack the blood supply, resist drug therapy, and cause the deterioration of healthy tissue, points out the severity of the disease and the imposing challenges of developing treatments.

Hereditary vs. Sporadic Cancer

We have seen that there is a close link between mutations and the development of cancer. In fact our current understanding of cancer suggests that all cancers arise as a result of cell mutations. When a cancer cell divides, it passes its mutations on to the two new cancer cells that are formed. However, saying that all cancers arise as a result of mutations in cells is not the same as saying that all cancers are hereditary. Mutations that arise in pancreatic cells may cause pancreatic cancer in that individual. But those mutations are in the cells of the pancreas; they are not in the egg or sperm cells that supply genetic information to the next generation. Egg and sperm cells are germ-line cells. Pancreatic cells are among the cells known as somatic, or body, cells. The genetic information from somatic cells is never passed to a new generation. So mutations in somatic cells may lead to cancer but these mutations cannot be inherited or cause cancer in the patient's children.

We do know that some forms of cancer can be inherited. A good example would be some forms of breast cancer. Breast cancer sometimes will "run in families" and the affected families will have a very high incidence of breast cancer. But even in these high-risk families not every woman is certain to develop breast cancer. What is inherited is a predisposition, or increased likelihood, of developing breast cancer. In light of what we know about cancer, what can explain this inherited predisposition to develop breast cancer? Well, we know that no type of cancer is the result of a single mutation. Each is the result of some series of mutations. Let us assume that we know that a certain series of mutations can lead to breast cancer. If an individual was to inherit from either parent, either through the egg or sperm, one of the mutations in the series, there would be one less mutation that would have to occur within the breast cells of that individual for cancer to arise. The mutation that they inherited from the parent would be present in all of the individual's cells, including the breast cells. Having inherited one of the mutations in the series increases the likelihood that breast cancer will develop, but does not make it inevitable.

Recently, much attention has been given to the discovery of "the genes for breast cancer." In fact what was discovered were two genes, BRCA1 and BRCA2. These are not the only two genes involved in the development of breast cancer. However, inheriting a mutated form of either of these two genes greatly increases a person's risk of developing breast cancer. It makes the most sense to refer to BRCA1 and BRCA2 as breast cancer susceptibility genes. In the United States approximately 1 in 12 women develops breast cancer. However, only between 5 and 10% of breast cancers are hereditary. Another way of saying this is that between 90 and 95% of breast cancer patients do not carry germ-line mutations that predispose them to developing the disease. In these patients, all of the mutations necessary to develop breast cancer occurred sporadically within breast cells some time during their lifetime. In the other 5 to 10% of brest cancer patients, those with hereditary cancer, at least one of the mutations necessary for cancer development was present at birth.

The BRCA1 and BRCA2 genes were discovered by examining families with very high incidence of early-onset breast cancer. The BRCA1 gene was mapped to a position on chromosome 17 and BRCA2 to a position on chromosome 13. Interestingly, even though BRCA1 was discovered in 1994 and BRCA2 was discovered in 1995, we are still not sure of the exact function of the proteins coded for by these two genes or how they specifically contribute to the development of breast cancer. BRCA1 is believed to be a tumor suppressor gene. Recent work on BRCA1 has suggested that it is involved in the damage control process, and may help recognize broken DNA molecules. If this is the case, it may explain why inheriting a mutation in this gene has such a strong effect on the likelihood of cancer development. If the defect from BRCA1 mutations reduces the detection of DNA damage, it would reduce the chance that it would be repaired. In these individuals, DNA damage would accumulate more rapidly and accelerate cancer formation.

The characterization of these genes points out another important aspect of cancer research. In families with inherited cancer, it is now possible to do genetic testing to determine just which individuals carry mutations in genes such as BRCA1 and BRCA2 that are known to increase the likelihood of cancer. These tests allow people with high levels of cancer risk to be identified. They can then go for checkups more frequently and improve the chances of detecting early any cancer that does arise.

SMOKING AND CANCER

Just as we all have a sense that a predisposition to some cancers seems to be inherited, we also all have a sense that environmental factors can be involved in the onset of cancer. A great example of this is the relationship between smoking and lung cancer. For a very long time, epidemiological evidence has been accumulating that links cigarette smoking to lung cancer. (Epidemiological evidence is simply data that draw a correlation between the occurrence of a disease and other factors.) As the prevalence of smoking in a population increases, the prevalence of lung cancer also increases. In the United States this was true when men began to smoke in large numbers and much later when women began to smoke in large numbers. There is also a relationship between the number of cigarettes smoked and the likelihood of developing cancer. The more cigarettes one smokes the more likely one is to develop cancer. Finally, stopping smoking can be demonstrated to decrease the likelihood of developing lung cancer. This is all strong evidence but none of it demonstrates a direct cause-and-effect relationship.

Just showing a correlation, or coincidence in the occurrence of two events, is not always completely convincing. For years the tobacco companies obscured the dangers of smoking by attacking the fact that it was "just" correlations between smoking and lung cancer that were used to argue that smoking actually causes cancer. It would be better to show cause and effect directly to demonstrate that smoking cigarettes can cause lung cancer. The first evidence for a direct cause-and-effect relationship came in 2001 when researchers were able to demonstrate that a chemical in cigarette smoke causes mutations in the tumor suppressor gene that codes for the p53 protein. The chemical, benzopyrene, is found in the tars of cigarette smoke. Ironically, in an attempt to rid the body of this insoluble chemical, the liver converts benzopyrene into the more chemically reactive benzopyrene diol epoxide. It is this chemical that interacts with DNA and cause mutations in specific sites in the p53 gene. As we have already learned, the p53 protein plays an important role in arresting cell division, activating repair enzymes, and if necessary, triggering cell death. The loss of these tumor-suppressor functions is an important step in the development of lung cancer.

The idea that cigarette smoke causes lung cancer by producing mutations also helps to explain another phenomenon. The link between cigarette smoking and lung cancer is not an immediate one. The prevalence of lung cancer in a population does not go up as soon as the rate of cigarette smoking begins to increase. There is a lag of between 20 and 30 years. We know that cancers develop through the accumulation of a series of mutations. It takes time for the entire series of mutations to build up, even if they are being produced in part by chemicals in smoke. Even if one or more of the mutations is inherited, the remainder of mutations must occur to ultimately lead to the category of disease we know as cancer.

Randall Phillis
Department of Biology
University of Massachusetts
Amherst, MA 01002
rphillis@bio.umass.edu

Steve Goodwin
Department of Microbiology
University of Massachusetts
Amherst, MA 01002
sgoodwin@microbio.umass.edu

Web Resources

Cancer Cell Cam
http://www.cellsalive.com/cam1.htm
This site presents images of human melanoma cells growing in cell culture. The sequence presents a number of cell divisions over a 24-hour period. A fresh image is loaded every 10 minutes. It is maintained on the cellsalive.com website authored by James A. Sullivan of Charlottesville, VA.

Cancer.gov
http://www.cancer.gov/cancer_information/
This site created by the National Cancer Institute (NCI) of the National Institutes of Health (NIH), contains straightforward information about cancer intended for patients, healthcare providers and the public. Of particular interest might be the section on clinical trials.

NCI
http://www.nci.nih.gov/
This is the official site of the National Cancer Institute, the primary U.S. government agency that oversees cancer research and treatment.

http://www.infobiogen.fr/services/chromcancer/
Atlas of Genetics and Cytogenetics in Oncology and Haematology
This site has resources for understanding chromosome rearrangements that occur in certain kinds of cancer, particularly leukemias. In addition, there are several links to other cancer related sites.

http://www.cancercare.org/
CANCERcare
This site has several resources related to cancer diagnosis, treatement and care of cancer patients.

http://www.nabco.org/
National Alliance of Breast Cancer Organizations
This site contains a comprehensive set of links and information about breast cancer.

http://www.fhcrc.org/
Fred Hutchinson Cancer Research Center
This is a world renowned cancer research center. There are a wide range or resources on this site including information for students about undergraduate and graduate opportunities to participate in projects and reseach underway at the center.

http://oncolink.upenn.edu/
OncoLink
This is the web site of the University of Pennsylvania Cancer Center. This has important links to information about clinical trials underway to test new cancer treatments.

http://www.cshl.org/public/overviews/cancer.html
Cold Spring Harbor Laboratory Cancer Research
This is the home page for cancer researchers at the Cold Spring Harbor Laboratory, a world leading institution for molecular biology research. Information from several labs working on current problems in cancer research are featured.

http://www.cancersource.com/
CancerSource
A rich array of resources about cancer types, and treatements.

http://www.academicpress.com/semcancer
Seminars in Cancer Biology—Academic Press
For cancer professionals, this journal has topical issues about specific cancer topics.

Books and Articles

Cooper, G. M. (1993). *The Cancer Book.* Jones and Bartlett Publishers, Boston.

Hanahan, D., and R. A. Weinberg (2000). "The Hallmarks of Cancer." *Cell, 100*:57–70.

Varmus, H., and R. A. Weinberg (1993). *Genes and the Biology of Cancer.* Scientific American Library (distributed by W. H. Freeman).

Weinberg, R. A. (1996). "How Cancer Arises." *Scientific American, 275*(3):62–70. (http://www.sciam.com/0996issue/0996weinberg.html)

Weinberg, R. A. (1996). *Racing to the Beginning of the Road: The Search for the Origin of Cancer.* Harmony Books.

Welsch, P., and M. C. King (2001). "BRCA1, BRCA2 and the genetics of breast and ovarian cancer." *Human Molecular Genetics, 10*:705–713.